The
Sawtooth
Wolves

The Sawtooth Wolves

Photographs by Jim Dutcher
Text by Richard Ballantine

Designed by Greg Simpson & Eric Baker

RUFUS PUBLICATIONS, INC.
BEARSVILLE, NEW YORK

THE SAWTOOTH WOLVES

Rufus Publications, Inc.

Text © 1996 Richard Ballantine

Photographs © 1996 Dutcher Film Productions

Compilation © 1996 Rufus Publications, Inc.

First Edition: September, 1996

Second Printing: Spring, 1997

Third Printing: Spring, 1998

Library of Congress Catalog Card Data:

The Sawtooth Wolves/Text by Richard Ballantine/Photography by Jim Dutcher

p.   cm.

ISBN: 0-9649915-0-0

1. The creation and life of a captive wolf pack.

2. Non-fiction. 3. Wolves. 4. Photography.

I. Nature and Natural History

96-83437 PCN

PROJECT DIRECTOR: RICHARD BALLANTINE

DESIGNERS: GREG SIMPSON & ERIC BAKER

PRODUCTION MANAGER: AMIE COOPER, THE ACTUALIZERS

SEPARATIONS: PIGUET GRAPHICS

PRINTED IN CANADA BY THE FRIESENS CORPORATION

DEDICATION

Pepe was a jaguarundi, or eyra, a wild cat from Ecuador sent to me by well-meaning but misguided friends as a surprise gift. Found from northern Mexico down to northern Argentina, *Felis yaguoaroundi* is the smallest, and according to some sources, one of the fiercest and most intractable of the New World cats. Pepe was just a few weeks old when he arrived in New York City, and easily fit into the palm of my hand. The trade in wild cats sold as pets is terribly cruel; 95% of these animals die, of various diseases but mostly from plain shock. Pepe was lucky. He took ill, I cared for him with all my might, and after days of force-feeding, he decided to live and pulled through. He eventually grew to a weight of perhaps 25 lbs., with a body about 24 inches long, and a tail of equal size. In my father's words, Pepe was to an ordinary cat "as a lightning bolt is to a candle." He could travel around a room without touching the floor, by running on the walls—up near the ceiling. Pepe had unbelievable strength, speed, and agility, and was utterly fearless. Yet he did not have a mean bone in his body. We played often and sometimes very vigorously, and never, not even once, did Pepe ever bite or scratch me. When Pepe was very little, he liked to nestle inside my shirt. When he was older, he would ride on my shoulder. At night he would sleep on the bed, and most mornings I would wake up with him curled around my neck. For a morning hello, he liked to lick my face, with a tongue that was like coarse-grit sandpaper. We went places together, visiting people or whatever, and often went out to my folks' house in the country where Pepe would play in the woods and splash in the stream. I scent-marked the boundaries of the property, and Pepe never crossed these lines. Pepe loved water and was a fine swimmer. In the off-season, we sometimes went to the beach and ocean. Pepe and I shared life and adventures and were happy together, but please do not misunderstand. Pepe was a wild animal. I was a young man living in a small apartment in a thoroughly urban city. It was not an easy combination, much less a fair one. We did our best. I stood by Pepe because he, an innocent, could not go back to the wild, and through no fault of his own was stuck with me-and because we loved each other very much.

PEPE BALLANTINE

WILD FRIEND
BORN
ECUADOR, 1964
DIED
BIG INDIAN, NY
1967

THIS ONE IS FOR YOU, PEPE — LOVE ALWAYS, RICHARD.

Contents

Introduction by Jim Dutcher

Jim Dutcher, with Kamots, a Northern Rocky Mountain gray wolf. Jim and his crew raised the members of the Sawtooth Pack from infancy so that the wolves would be at ease around humans and their natural behaviors could be filmed. "Wolf: Return of a Legend" won an Emmy Award and has been seen by over 200 million viewers world wide.

Wolves are curious and playful, and initially, setting up the cameras would often distract the wolves, who would stop whatever they were doing to investigate the filming equipment. After a while, however, the wolves became accustomed to the cameras, and would go about their business as usual. Below: Pups at 14 weeks with Jamie Dutcher.

The publication of this book marks the close of a wonderful project that has enriched the last six years of my life. With my dedicated crew, we have had the privilege of being accepted by a family of wolves and allowed to be part of their world. 🐾 As an endangered species, wolves are protected from humans. Photographing this threatened animal in the wild (if at all possible) could cause major disturbances. In some cases it would not be allowed by the U.S. Fish and Wildlife Service. Therefore the only way to film this shy and elusive animal without interfering with its well-being was to form a captive pack. The approach I took with this project was designed to gain the trust of the wolves so that I could be amongst them to film. The face to face images I obtained are due to my bond with these wolves that made them comfortable with me. 🐾 We committed time and constructed the largest wolf enclosure in the world. Our pack was assembled from the offspring of captive wolves and the pups were bottle fed by us as surrogate parents. However, the wolves of the Sawtooth Pack are not pets. They are not tame animals. They are familiar with people but behave as a pack of wolves with their own rules. The unique bond we developed has allowed us to observe them in a relaxed and natural state, as opposed to observing a wild, non-socialized pack that is always on guard around the scent and sight of humans. These wolves may even consider us human pack members. We will never know for sure. 🐾 In order to observe the wolves without interfering with their

> *"The quickest way to learn the language of a species is to do so as a social partner."*
>
> — KONRAD LORENZ

natural behavior we eventually constructed a tented camp in the middle of the enclosure. Surrounded by fencing, we built a platform seven feet high and perched a Mongolian-style yurt on it. We cooked and ate our meals in this warm, round tent which also served as a sleeping facility for one or two crew members. Below this platform, my wife Jamie and I slept in a wall tent crowded with filming and sound recording equipment. Often at night the wolves would howl into the absolutely silent, dark sky and Jamie, keeping her sound gear warm in the sleeping bag, would record their songs. ❧ Both the yurt and the tent held wood stoves which we kept burning all day and into evenings during the bitter cold months of winter. However, the fire would go out while we slept and the condensation from our breath formed a layer of ice on our pillows that proved to be a rude awakening as we turned over in our sleep. By committee decision it turned out to be MY job, not Jamie's, to brave the cold floor in the mornings and re-light the fire to defrost our clothes for the day. Our unique view of wolf life came from living within the pack's territory and the acceptance of our presence was amazing. Not only could we watch, listen, and film while respecting their inter-group social structure, but they would also solicit our attention and affection when we left our compound to join them. We soon began to see the wolf as an animal that is capable of being caring, gentle, courageous, and playful. These were the behaviors of animals who are intelligent and social.

The wolves approach Jim and the crew on their own terms.

Jim's film "Cougar: Ghost of the Rockies" aired on ABC-TV in 1990, and featured a mother cougar and three new-born kittens. Cougars can be unpredictable and very dangerous; filming close-up required great care and sensitivity from Jim and his crew. Below: Cougar kitten that was eventually released into the wilds of Idaho.

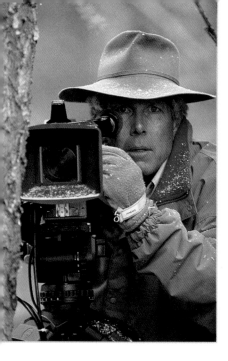

Wildlife cinematographer, Jim Dutcher, began this project in 1990. He has formed a bond with the Sawtooth Pack that is unrivaled.

Photographer and sound recordist, Jamie Dutcher has a background in animal husbandry. She contributed a number of photographs for this book and has played an instrumental role in the Sawtooth project.

Displays of dominance and submission took place frequently, reaffirming the social balance and reminding each wolf of its proper place in the pack. Often Jamie and I would be caught in the middle of one of these disputes, surrounded by growling and yelping wolves as the more dominant animals reinforced their positions in the pack hierarchy. Kamots, who finally emerged as the pack leader, would then end the fight with a howl before any harm was done. Following the alpha's lead, all members of the pack would join in, literally howling all around us, rallying together to show solidarity, licking each other's faces and ours as well, and then whirling away to be together—creating a sense of their bonds but also of their wildness. 🐾 Our project deviated from conventional filmmaking. The easy part was knowing, within about 25 acres, where our subject would be every day. (We hoped!) What surprised us was that we were dealing not only with the traditional process of filming such as the selection of lenses, filters, film and

other camera gear, but with a tremendous number of new problems as well. State and federal government permits and regulations needed to be met, ranging from endangered species to public land use, from firewood to the specifications of the enclosure. 🐾 We were also concerned with other predators such as cougars and bears in the area and whether they would respect the fencing. Forest fires burned in the Sawtooths during several of the summers and as Wolf Camp filled with burning cinders, our fears for the pack escalated as the fires encroached. We needed to find roadkilled deer, elk, and antelope, haul the roadkill in eighty-degree heat or minus

One of the most vital winter jobs at Wolf Camp; shoveling the path to the outhouse.

thirty-degree cold to store it in a freezer down the valley, haul it out and feed the pack. Most importantly, our concerns revolved around the constant responsibilities and worries of keeping these captive wolves happy and healthy. ❧ At 6,500 feet above sea level, our camp experienced long winters, with snow piling up to depths of four to six feet. During the short days of winter the sun would set behind the 10,000-foot peaks by 2:00 PM. Snow did not fall just in winter. Even throughout the summer months we had surprise snow showers. ❧ Our enclosure had to be inaccessible to maintain the behavioral "intactness" of the wolves, for research, and for filming. This necessary remoteness presented hardships, especially in winter. We had daily worries of maintaining the enclosure and the needs of the animals in a demanding environment. Would this windstorm bring a tree down across the fence?

The yurt was at first outside the enclosure. Later, it was moved inside, on top of a 360-degree observation deck that made it easier to observe and film the wolves without disturbing them.

Opposite page: Wolf Camp. Located 6,500 feet high in the Sawtooth Mountains near Stanley, Idaho, the camp included open fields, a stream, thick undergrowth, and forest. There was plenty of room for the wolves to be by themselves.

W ould this snowstorm be so deep that the wolves could jump the fence? Would the trail to camp be visible in the whiteout blizzard? Would the snowmobiles even start at twenty-five below? We dealt with equipment problems, wood stoves that would not light, deep snow that immobilized the crew, ice, frozen toes, and cold outhouses. At the same time we were surrounded by awe-inspiring beauty with the Sawtooth Mountains rising cold and huge behind camp, and the wolves at home in winter, curled up to sleep in minus 40 degree nights. 🐾 I grew up with a love of nature. Raised in southern Florida, I spent my early years around, on or under the ocean. I was scuba diving by the time I was eight and underwater photography followed shortly. My parents gave me an underwater 16MM movie camera for my high school graduation. I completed my first film at twenty-three and when it started winning awards around the world, it pointed a direction for my career and I was hooked. 🐾 I produced a PBS Special called, "Water, Birth, the Planet Earth" that followed the cycle of water from the Earth's atmosphere through the river systems to the sea. Without realizing it, the filming of this project moved me away from my familiar underwater world and opened my awareness to the wildlife of the Rocky Mountains. 🐾 My first project in Idaho was the filming of "Beaver Pond" for National Geographic. By nature shy and nocturnal, filming beavers presented many challenges. In order to explain their secret life, we constructed a beaver den inside a cabin, with built-in viewing windows for filming. We built a tunnel through the wall so the beavers could go outside to swim. The beavers accepted these living quarters and we were able to film them over a two year period in the den and underwater. We duplicated their home so well that the beavers relaxed and carried on their routine to the point of mating. This was the first time a beaver birth had been caught on film. It was quite an ordeal. Not knowing when they would give birth, our crew of three stood watch in alternate shifts, observing the beavers every twenty minutes for thirty days to capture that moment on film. 🐾 The scope of our projects started to get bigger. Carrying on this theme of taking an audience into a world they had never seen before, we began plans to film an animal seldom observed: the cougar. All our thought and effort went into finding the proper place and designing an enclosure with enough habitat to allow the animals to live as normally as possible. With special permission from the Forest Service, we enclosed five acres of public lands. Working with wildlife researcher Maurice Hornocker, we acquired a pregnant female cougar who, with her soon-to-be-born three kittens, became the focus of "Cougar: Ghost of the Rockies" which aired on ABC in 1990. We took great care not to socialize the kittens with human contact, and with their mother's training, they were ready for a normal life when we released them on a remote mountaintop, wild and free, at the conclusion of the project. It felt good to give something back to nature.

Searching for a new subject, I realized that the most misunderstood and wrongly feared animal in North America was the wolf. Researching the wolf, we quickly learned of its notorious reputation. Evil. Vicious. Bloodthirsty. However, no human has ever actually been killed by a healthy wild wolf in the history of North America. We wanted to dispel the myths and create a new understanding of wolves. The continual negative and extreme reactions I faced in setting up this wolf project confirmed my feelings that this was a story that needed to be told. ❧ All of my film projects have taken a lot of time, a good chunk of my life, and connections with the animals have developed. But none so strong as with these wolves. I will never forget my time spent with them. After all, I was there bottle-feeding them when they were puppies. What wonderful, rich memories I have—the joy they seem to express when I return to camp or the howls when I depart. I actually believe they will miss me as much as I miss them. Our bond is that strong. ❧ Since the inception of this project, we always knew that these captive wolves could never be set free. Working with the Nez Perce Tribe and the Wolf Education and Research Center, a plan was devised to build an enclosure on Tribal lands so the wolves could continue in their role as ambassadors for their kind, to educate the public as to the true nature of these misunderstood animals. The new enclosure, as big as the last, is designed to allow the pack space to run, play, explore, and be alone. ❧ While there will be sadness in saying farewell to these wolves, in seeing them leave this wonderful location, and in their entering an unknown place, I know the Tribe's welcome will be a reassurance to both the pack and myself. ❧ When I look back over the long winters of sub-zero temperatures and frozen fingers trying to operate a camera, I may shiver for a moment. But this experience has provided a warmth that will last a lifetime. I hope this book will convey those same feelings to you.

— JIM DUTCHER

Wahots

The Project

Top: Kamots; middle: Makuyi; above: Akai.

The Sawtooth Wolf Pack started in the summer of 1991 with two adults and four pups. Akai, the adult male, and Makuyi, the adult female, who had not met before, were placed in the main enclosure, to explore their new home and to get to know each other. The pups, Kamots and Lakota, brothers in the same litter, their sister Aipuyi, and Motaki, a female from a different litter, were eventually placed in a pen at the camp with a dark cardboard box to serve as a surrogate den where Jim and his crew would raise them to accept humans without fear or threat. 🐾 Socializing wolves to humans is an elaborate and intensive commitment. Jim and his crew were the parents for the wolves. The pups looked to them for food and contact, and as dutiful parents one of them would rise out of a warm bed at 6:00 AM to walk across an icy floor, stoke the fire, warm specially prepared milk, don long pants and a long sleeve shirt, and enter the pen to feed the pups. With sharp claws, needle-like teeth, and an aggressiveness that has evolved to prepare wolves for a very demanding world, the pups would energetically compete for the bottles. This early physical contact was a vital part of their socialization process. Wolves develop and grow quickly, and soon the

pups were becoming independent, playing more aggressively, and beginning to explore their surroundings. Play is a central activity in a wolf pack at every stage, but pups especially use play to develop strategies for social domination and submission, the hunt, flight, and to develop muscles and reflexes. ❧ Puppies form their own hierarchy, wrestling their way into a rank order of their own making. Far from gentle by human standards, always restless and in motion, the pups were a challenge to handle and socialize while they established their personalities and began to vie for social status. ❧ Occasionally Jim or one of his crew would have to restrain or discipline the puppies, and did so with careful regard for the language of wolves and the traditions of wolf society. It was usually sufficient to roll a pup over into the belly-up submissive position. Once in a while it was necessary, exactly as a wolf parent would do, to cause a pup to pause by issuing a low-pitched, warning growl. ❧ Jim and the crew were firm at times with the puppies, but even the idea of a physical confrontation was avoided. Rather, they treated the wolves as they wanted the wolves to treat them: gently, and with fairness and respect.

Top: Akai exploring his new home; above: Akai (left) and Makuyi meet. Below: Akai; bottom: Makuyi.

When they began to howl, the puppies were taken out of the tent that served as their den into a half-acre fenced enclosure where tall grasses, trees, and a creek became a universe of endless curiosity and smells. The puppies continued to play and wrestle with each other, building up their strength, coordination and agility, while learning skills and sorting out their own hierarchy, but they also increasingly went off investigating on their own—and nearby was something of great interest. The main enclosure with Akai and Makuyi in it was only a few meters away from the puppy enclosure. Over the course of the summer, by sight, smell, and sound, the adult wolves and puppies exchanged a great deal of information about each other. The eyesight of wolves is about on a par with that of humans, but their senses of smell and hearing are extraordinary. The pups and the adults were familiar with each other long before they were formally introduced. 🐾 Jim and the crew continued to nurture and raise the puppies for about four months. This extended period of socialization was the crucial stage in the formation of the Sawtooth Wolf Pack, the wolves bonded with the humans: Kamots, Lakota, Aipuyi, and Motaki accepted Jim and his crew as part of their family.

Right from the start, the pups gave signs of their personalities and future pack roles. Kamots, who became pack leader, was confident, unafraid, and curious, and could be taken from the litter without becoming upset. In contrast, Aipuyi, who was always vying for status with her packmates, was uncomfortable when being held. Motaki, who became the first omega, was deeply subordinate and timid, and would hide when humans approached. Lakota was quiet, distant, and wary with watchful intelligence.

There are two main species of wolves in North America: the red wolf (*Canis rufus*), and the gray wolf (*Canis lupus*). The red wolf is limited to the southeastern United States. The gray wolf ranges from the area around the U.S./Canadian border up almost all the way to the North Pole. ❧ At one time, twenty-four subspecies of North American gray wolves were recognized, based on regional variations in overall size, color, and skull configuration. However, wolves are travellers, and the movement of wolves across regional borders and interbreeding between subspecies has blurred the distinctions between types. Today, many scientists divide gray wolves into five subspecies, and names such as eastern timber wolf and northern Rocky Mountain describe geographic origin rather than physical characteristics (just as a person from Colorado might be short or tall). Gray wolves are citizens of the world, and basically are all much the same. The physical differences between them relate to food and climate. ❧ An average-size North American male gray wolf might weigh 95 to 100 pounds, stand 30 to 32 inches at the shoulder, and stretch six feet from nose to tail (females are about 20 percent smaller). A big wolf might weigh 120 pounds and stretch six and a half feet. ❧ The largest wolves are found in mid-Canada and Alaska, where body weights up to 175

Bonding in a Wild Wolf Pack

A wolf pack is an extended, closely-knit family unit bound by intense care-giving between its members. The future of the pack is in the young, and a wolf birth is an event. Shortly before a birthing the pack will gather outside the mother's den in a state of high excitement, scratching at the ground, whining, and howling. Yet no one, not even the father, may enter the den when the pups are being born. ❧ After the birth the pups stay inside the den for the first three to four weeks of their lives where they are looked after and nursed by their mother. The father and sometimes other pack members bring food to the mother. At about four weeks the pups begin to go outside the den, and at five to eight weeks they are weaned. Feeding the pups is then the responsibility of all the pack members, who regurgitate food for them. If the mother joins a hunt, other pack members stay behind to care for and look after the pups. ❧ At about eight weeks, the pups are moved from the den to a nearby rendezvous site, usually near water, where they can explore and play and be with the pack. This is the initial learning stage, and the adult wolves, who in general express a strong attraction and interest in pups, are remarkably tolerant as the youngsters climb over, under, and around, and pull and chew on them. ❧ At about twelve weeks the pups start to accompany the pack on hunting trips. This phase is a new learning phase, the business is survival, so the training is specific and firm. All members of the pack participate in teaching the pups, helping or disciplining them as necessary. Learning and growth are rapid; at seven to eight months the pups are actively hunting, and have attained most of their full size. ❧ The care, attention, and training lavished by adult wolves on their young makes the pack an efficient hunting unit that is cohesive in identity and territory, it also forms enduring bonds of intimacy, affection, and friendship between individuals, one and the same with the pack.

The pack has just fed on a road kill, and are carrying away scraps that they will later eat, or use for play.

pounds have been reported. Arctic wolves are lighter, and have tall, rangy bodies for covering long distances. In contrast, the wolves that once lived in hot areas such as the U.S. southwest and central Mexico, were compact and small, and weighed 50 pounds or less. ❄ Gray wolves come in all shades of gray, tan, brown, rusty red, cream, and buff, and solid black or white. Most gray wolves have a silver-gray body and light tan legs. About 30 per cent are black, with this color most prominent in Alaska and mid-Canada. Arctic gray wolves tend to be creamy white. The fur coat is comprised of two layers: the outer guard hairs shed water and provide the prelage or markings, while the dense undercoat provides insulation. A wolf can curl up in the snow and sleep at minus 50°F. ❄ Wolves have large territories and are built for distance running. The chest is narrow, which makes forging through snow easier, the legs are long and closely set together at the front, so that one follows the other in the same track, and the paws are large, almost the size of a man's hand, for traction on snow. ❄ As befits top-line carnivores, wolves have large teeth, and jaws with a bite force of 1,500 pounds per square inch, capable of crushing the thigh bone of a moose.

In the late summer the pups were introduced to Akai and Makuyi. Akai, the male, was immediately comfortable with the pups, and took them on a tour of the enclosure, using the occasion to firmly establish his position as leader. Makuyi, on the other hand, became confused and frightened, suddenly bolting in full flight from the group, to hide within the thick undergrowth at the far end of the enclosure. No inducement would make her come out.

Akai with pups.

As fall became winter and the snows fell, the pups grew and flourished, at every opportunity following Akai and playing with each other. Akai appeared to take on responsibility for their care. The pups were now juveniles, physically nearly full-grown, powerful and fast, and often vigorous and rough with each other. This was the serious period for them of finding their place in the pack hierarchy.

Makuyi at the end of the enclosure, never integrated into the pack.

Wolves are always intensely curious about new things, and have a particular fascination with objects. Anything inside the enclosure was fair game: items of clothing, cameras, tools, even the sleds used to transport equipment and food. The crew had to be careful not to leave anything unattended, or else one of the wolves would be sure to snatch it and run off with it.

Kamots leading a group howl.

Makuyi with food.

Meanwhile, Makuyi continued to live in hiding at the far end of the enclosure. Jim regularly brought her food and spent time with her, eventually creating a gentle bond of friendship and trust. Makuyi was quiet and friendly and this made her self-imposed isolation all the more puzzling. One day Jim noticed that Makuyi's eyes were clouded and that she did not seem to see well. He began to suspect that this might be the source of her fear and confusion, and of her unwillingness to be with the pack.

One cold morning in February, Makuyi had a change of heart. She came out of hiding, approached the juveniles, and tried to bond with them. Makuyi was in estrus (sexual heat), which occurs only once a year for a wolf, and her need for companionship had overcome her fear. Instead of welcoming her, the pack rejected Makuyi, chasing and biting her whenever she came near.

Jim and Makuyi.

Makuyi.

Wolves use their eyes a great deal in communicating with each other. If Makuyi was really having problems with her eyesight, as Jim suspected, this could be the reason for her rejection. A veterinarian was called in and confirmed that Makuyi had cataracts and could hardly see. Jim decided on an extraordinary measure: Makuyi would have an operation, right at Wolf Camp, to restore her sight.

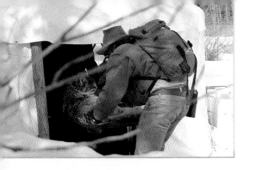

Wolf Eyes

Wolf eyesight is attuned to hunting. The eye has no foveal pit, a depression at the back of the eyeball that enables sharp long-distance focusing. As a result, wolves are not good at seeing things that are far away and motionless. However, the outer edge of the retina is highly sensitive to movement, so wolves have excellent peripheral vision and quickly notice motion. As well, the retina has a high ratio of rods to cones, predominantly the red type, which helps wolves to see well at night. A further aid for nocturnal vision may be provided by a light-reflecting layer called the *tapetum ludidum*, which also makes wolves' eyes appear to glow in the dark. 🐾 How wolves see is not just a matter of physical capability. Wolves live in an ongoing state of high general awareness and watchfulness, and notice and assess small details quickly, because this is essential for survival.

he operation was performed by Dr. Grant Mauer, a veterinary ophthalmologist, and Dr. Randy Acker. Travel would have been an additional stress for Makuyi, so the tent was transformed into an operating theater, with a portable generator for power. Makuyi was given a tranquilizer in her food. After she fell asleep, she was carried into the tent and placed on the operating table. When she was examined, a number of severe bite wounds were found hidden underneath her thick winter coat. These were extensive and had to be attended to before the eye operation could proceed. When Dr. Mauer examined Makuyi's right eye

he found that the lens was completely opaque. She could see nothing with that eye. Dr. Mauer removed the damaged lens and replaced it with a synthetic one. This operation had never been performed on a wolf, but to everyone's relief, after two months of care, Makuyi totally recovered. Makuyi was placed in a separate enclosure together with three new pups being raised to join the pack. Makuyi happily cared for the pups, but an attempt to re-introduce her into the Sawtooth Pack did not work; too much strife had passed, and the pack continued to reject her. Makuyi was returned to Montana to live with her siblings.

Eyes as Soul

Wolf eyes are blue at birth and then change to a yellowish-amber color. A wolf's gaze is striking. The iris has a characteristic luminous glow which is heightened by the black of the pupil, and frame of black eyelids. By the eyes, a wolf appears at once intense and utterly alert, yet completely relaxed and neutral. Wolves are careful about how they use their eyes, and stare at each other only in prelude to combat or play, when small changes in the size of the pupil and other clues can indicate changes of mood and signal action. Wolves express confidence by looking to one side or "through" another wolf. Makuyi's blindness was not just a defect, but a major block to communication, and this may have been one reason why the pack rejected her.

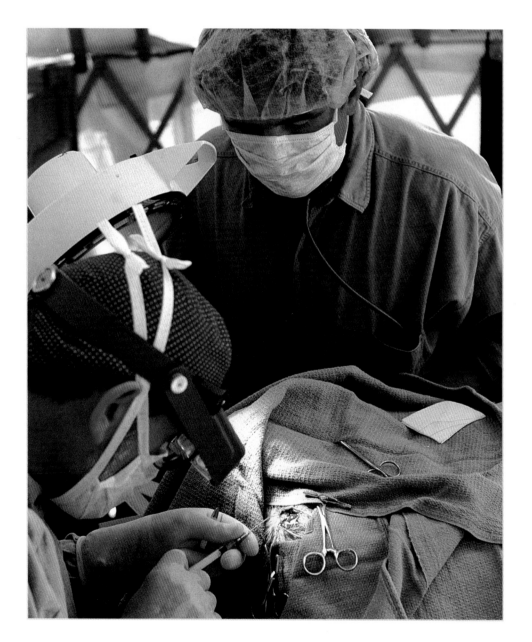

Top: Motaki and Kamots during a quiet moment; above: Motaki, the omega, initiates play with Kamots, the leader.

Above: Kamots gives chase to Motaki; below: Jim discovers Motaki has been killed by a mountain lion.

Meanwhile, the pack continued to develop its social structure. As the only adult male, Akai was, by default, pack leader. Although he never completely took on the alpha command role, the younger wolves did not challenge his natural place as head of the pack. Among the juveniles, Kamots, confident and strong, was clearly the leader. Aipuyi, his sister, was unpredictable: sometimes cheerful and playful, other times temperamental. Lakota was calm, aloof and reserved. Motaki, submissive and fearful, was the omega, the pack's scapegoat. At this stage, however, it was actually Motaki who made the greatest contribution to the functional development of the pack. Motaki was the chief instigator of play. She loved games, and she used humor to deflect abuse away from herself, as well as to break apart unruly conflicts between other pack members. The job of an alpha is to be alert to possible outside threats to the pack, and to make decisions. As omega, Motaki was more involved in the emotional lives of her packmates, and so, in some ways, it was Motaki who looked after the well-being of the pack. She was submissive, but that did not mean a loss of dignity, or of caring. Her role made her an active and important member of the pack. One morning in mid-June, Jim made a terrible discovery: Motaki was dead, killed by a mountain lion during the night. Why did a cougar climb a high fence to enter the enclosure? Did the pack join the fight? No one will ever know the answers to these questions. For both Jim and the Sawtooth wolves, exactly how or why it had happened was completely eclipsed by sadness and shock: Motaki was dead. All play in the pack stopped for six weeks. When the pack passed through the aspen grove where Motaki was killed, the wolves would become noticeably transformed in a quiet and depressed way. And when the pack howled, Jim heard a new, eerie quality in the wolves' voices that sent chills down his spine. The pack was in mourning.

Gentle and sweet-natured, Motaki was sensitive to her packmates. After she was killed by a mountain lion, the rest of the pack seemed to mourn her loss deeply.

Kamots with Jim (top). Wahots with Jamie (above and opposite).

Feelings in Animals

Do animals have feelings? Can wolves experience emotions such as happiness, joy, sadness, and bereavement? One way to answer this question is to ask: do we have feelings? Of all animals, wolves are very near in nature to humans. Most animal researchers agree that wolves have complex feelings. The difficulty is in identifying what those feelings are. ❧ Given their similar family dynamics, wolves and humans doubtless share some feelings; given their dissimilar lifestyles, other feelings are probably unique to each species. Another complication is that wolves, like humans, are individuals, each to be taken on their own terms. ❧ In any case, wolves and humans both operate strongly in terms of feelings. It is through our feelings that we relate to and experience the feelings of wolves, and most certainly, it is through their feelings that wolves experience us. People who live and work with wolves are well aware that wolves are highly sensitive to emotions, both within themselves, and in humans. Feelings and emotions are subjective and prone to change, and therefore are not easy to understand and describe, but for the wolf they are the stuff of life.

Matsi	Matsi
Motomo	Amani
Motomo & Amani	Amani

Akai never bonded with Jim or the crew, and a change in his temperament was unlikely; even long periods of socialization may never be enough to create a bond between a wolf and a human who has not raised it. At the end of August, 1992, the decision was made to move Akai to a new home at Mission: Wolf, a sanctuary high in the Colorado mountains, where he could live with a female wolf. 🐾 With Akai gone, Kamots, who was now 15 months old, immediately assumed the role of alpha pack leader. Tail high, head and ears forward, eyes alert, Kamots was vigilant in protecting the territory of the pack. He was the first to howl, to summon the pack, and the first to the carcass when the pack was fed a road kill. His confident, benevolent nature was invaluable when, in mid-September, three new pups were introduced into the pack. 🐾 Motomo, Matsi, and Amani, all brothers, had been living in the puppy enclosure for some time. Motomo, jet black and with intense yellow eyes, was filled with curiosity and always the first of his litter to investigate something new. He liked attention and got it through a combination of persistence and a talent for mischief. Matsi was uncommon looking and very beautiful, with a gold rather than gray mask around umber-colored eyes, and an elegantly-structured face with a long snout. Gentle and neutral with the other wolves, Matsi formed a special relationship with Kamots, and the two often slept next to one another with their forelegs intertwined. Amani had the darkest face mask of all the wolves, set off by deep gold eyes. Mild-mannered and liking attention, Amani preferred eating and playing to jostling for rank within the pack. 🐾 In a somewhat surprising development, Lakota, usually distant and reserved, took on the job of caring for the new pups. Patient and forbearing in his puppy sitting duties, Lakota also became one of the most playful members of the pack, often leading the others, especially Matsi, in games of chase.

Wolf pups are born in the spring. A wolf pup weighs 1 lb at birth, and grows rapidly, on average gaining 3 lbs per week for the first 14 weeks, and 1.3 lbs per week for the next 14 weeks, reaching nearly full size in time to be able to run with the pack in winter.

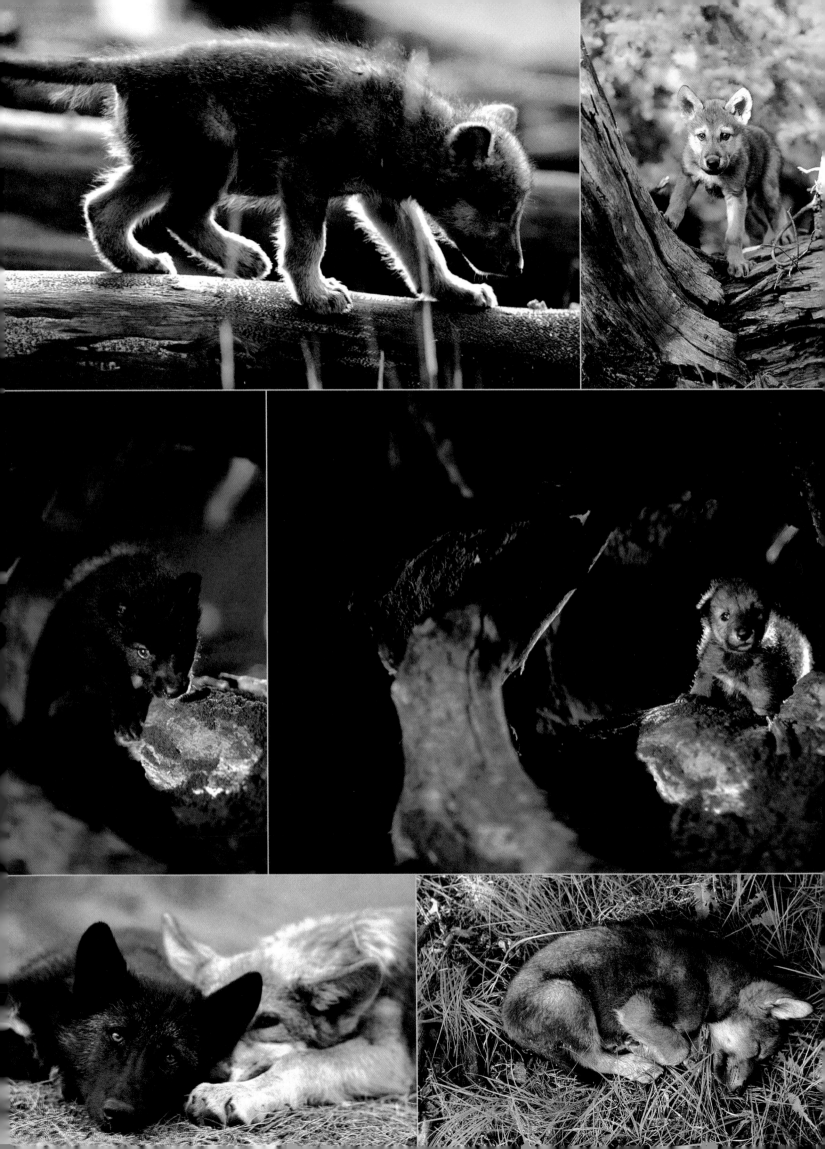

Matsi	Lakota, Motaki, Aupuyi
Aipuyi and Motomo	Matsi and Motomo
Kamots and Lakota	
Lakota	Kamots and Lakota

Opposite page: Kamots sleeping on a bed of willow branches. In snow, wolves sometimes make sleep sinks from branches, leaves, or the leftover hair of a deer to provide additional insulation. A wolf can comfortably sleep outside with the temperature at fifty degrees below zero F.

Aipuyi became the pack omega. Aipuyi, however, had none of Motaki's tolerance nor her sense of play. Far more aggressive, Aipuyi refused to retreat from conflicts and instead escalated incidents. She became involved in an increasing number of fights with younger pack members and the crew noticed prolonged social tension and increased anxiety in the pack. Finally, after a fight that was near-fatal, Aipuyi was removed from the pack. Due to

Top: Aipuyi and Lakota playing; right: Aipuyi attacking Motomo.

the severity of her injuries, it was necessary to put her to sleep. 🐾 After Aipuyi's departure, the pack settled into a new equilibrium and an easier time with increased play and less aggression. Kamots became the alpha leader, and Matsi the beta, or second leader. Wary Lakota, whose habit of careful observation suggested keen intelligence, became the omega, often leading the pack in play.

Kamots, the alpha, is the largest wolf in the pack. He is quick to demonstrate his physical superiority if challenged, but his leadership rests primarily on his qualities of alertness, confidence, and even temperament.

The Sawtooth Pack 1992

KAMOTS

Alpha Leader

REGAL, CONFIDENT AND BENEVOLENT.

LAKOTA

Omega

INTELLIGENT AND CARING.

MOTOMO

Middle Rank

BRIGHT, CURIOUS AND ENERGETIC, OFTEN INITIATES PLAY.

MATSI

Beta

SECOND IN COMMAND, A PEACE-MAKER.

AMANI

Middle Rank

MILD-MANNERED, GENTLE AND FRIENDLY.

Family Life

Top: Kamots; above: Lakota and Motaki were both at different times in the omega position within the pack; below: Matsi submitting to Kamots.

Like humans, wolves evolved as cooperative family groups. Members of any healthy family develop specific roles so that the entire family functions more efficiently. Decision making is shared by wolves, and although traditionally it has been thought that the alpha or lead wolf determines who will eat first, mate, and hunt, there is great flexibility in role sharing. ❧ Wolves subsist on small game such as chipmunks and rabbits, and by hunting large, fast animals—deer, caribou, moose, bison, and elk—that can be up to ten times heavier and more powerful than an individual wolf. Of necessity, wolves are strong, hardy, intelligent and adaptable. Working as a team gives them greater efficiency when hunting. The pack, however, is far more than a simple mechanism for survival. ❧ The dynamics of a wolf pack are complex, intricate, and profound. Interaction between pack members transmits past experience, regulates individual behavior and pack structure, and keeps the pack at a size that is ecologically balanced within the extent of available prey. ❧ The ability of an individual wolf to put aside his or her immediate needs for the greater good of the pack—for example, to go hungry so that a pup may eat—defines a wolf pack member. Equally, it is the ability of the pack to recognize and meet the needs of each individual wolf within the group that determines the diversity and hence the strength of the pack. As Kipling astutely wrote over 100 years ago in *The Jungle Book* (1894), the strength of the pack is the wolf, and the strength of the wolf is the pack. ❧ A wolf pack is organized as a hierarchy, with a clear rank and role for each member. Much of the pack's activities are directed by the leaders, the alpha male and

female, but the pack also works as a fluid unit, with wolves of all ranks contributing to the group's social direction. ❧ A wolf pack is in many ways like a family, and feelings, emotions, and personality traits are essential elements in the inter-dependence of the pack. Playfulness, humor, jealousy, irritability, benevolence, tolerance, depression, rage, celebration, joy and other emotions allow wolves the flexibility to respond to social change, danger, and seasonal occurrences. ❧ The behavior of wolves is variable, and at any given moment, can change depending on their mood, recent events within the pack, and environmental factors such as the weather, availability of food, and time of year. When an individual wolf asserts his or her status, or attempts to explore the limits of behavior acceptable to more dominant members of the pack, the challenge may be accepted and turned into a playful situation, or it may be severely chastised. ❧ On a warm afternoon with the Sawtooth Pack peacefully lazing in the sun Amani may get up, stretch, and then stand over Lakota, growling. Lakota may, after submitting to Amani, walk over to Kamots and face-lick him. Kamots might respond with a gentle gesture—or with severe dominating postures. ❧ Over the years that Jim and the crew lived together with the Sawtooth wolves, they found that the social organization of the pack has intricate nuances. Events and communications between the wolves have subtle, myriad permutations which are evident only with time and familiarity. The reactions and interactions of the wolves are always to some degree unpredictable, because their family life is rich and complex, and the wolves are constantly changing and evolving.

Above: Kamots soliciting play from the pup, Matsi; below: Matsi as an adult; below next: Kamots disciplining the young Motomo; bottom: Chemukh in submission to Kamots, is trying to win her way into a feeding place on the carcass.

Diagram of Wolf Pack

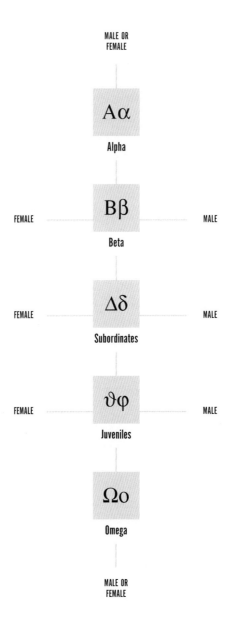

Hierarchy

Awild wolf pack is basically a family, composed of a breeding pair, the alpha male, alpha female, their offspring, and perhaps a few distant relatives or even a wolf or two from another family, if they have a useful function within the pack. A pack can number from 2 to 30 wolves. Most packs in the U.S. number from 5 to 10, in Canada from 10 to 20. Larger packs are more efficient at hunting big animals such as moose. 🐾 The pack hierarchy has two lines of dominance, male and female. The alpha male dominates the second-rank male, the beta, who in turn dominates the third-rank male, and so on down to the bottom-rank wolf, the omega, who may be male or female. The female line is similarly ordered. Pups are subordinate to juveniles, but are neither above nor below the omega. ❧ The alpha pair control food and reproduction. The alphas eat together, and direct the order in which the rest of the pack feed at a kill. Typically, the alpha female is the exclusive breeder in the pack, and during the mating season vigorously prevents the other female wolves from mating. In rare instances, if food is abundant, or if the pack has been unbalanced by deaths caused by human hunters or other causes, a beta female may produce a litter at the same time as the alpha female. ❧ Traditionally, the alpha male is supreme leader of the pack, but alpha females also lead packs, and alpha pairs can share and interchange roles. Whichever is in command, the qualities of the alpha leader are important to the general tenor of the pack; packs led by confident, benevolent wolves are more harmonious and efficient than packs led by nervous, repressive wolves. ❧ The individual natures of wolves affects the success of their roles within a pack, and at the same time, wolves can and do change their natures to suit different roles. A lordly alpha made infirm by age and deposed to the status of omega will of necessity become fawning and submissive. The alternative is banishment, or even death. The pack is that important. A scrappy, pugnacious beta who has maintained position by aggressively dominating lesser rank pack members may, upon graduating to alpha status, become more considerate and altruistic. It is a question of playing a part, of living up to a role that every wolf in the pack understands. ❧ Wolves also express their individual natures and alter the roles they play by forming alliances and special relationships with each other that transcend the pack hierarchy. A beta and an omega normally at odds when feeding may be close friends when off by themselves away from the pack. They may even reverse roles. ❧ Wolves live with, and adjust to, changing conditions and circumstances. As time passes in a wolf family, members grow old and die, new ones are born, become sexually mature, and otherwise undergo changes. The wolves' understanding of the pack hierarchy and roles, the alliances and friendships they form, and the emotions they feel and express, all serve to maintain both the pack, and the individual wolf. The business of the pack is not just to eat, but to ensure that it endures over time. It has been so for a million years and more; always and forever, the wolf is the pack, and the pack is the wolf.

Alpha

Alpha wolves are the dominant pack leaders. The alpha male and female are the decision-makers. They are also the first to deal with outside threats to the pack, and are the controllers of the social structure within the pack. 🐾 Puppies form and maintain their own hierarchy within the pack for the first year or two and positions are found early. In the Sawtooth Pack's first group of pups, Kamots' personality left no doubt as to who was the leader. In his position of alpha male, Kamots usually initiates the howl. When he does, the rest of the pack comes to join him.

Top: Kamots initiates a howl; center right, bottom left and right: Lakota and Motomo join Kamots; opposite: Kamots defends a carcass.

Omega

Opposite from the alpha is the omega, bottom-rank in the pack hierarchy. Unlike the beta position, which occurs in both the male and female lines of dominance, there is only one omega, who may be male or female. It is thought that the hardship of being picked on by other pack members may promote resourcefulness in omega animals, making them more adept at diverting and diffusing tensions and conflicts through games and play. Nonetheless, an omega must withstand a lot of physical abuse. After an attack, Lakota may face-lick, whine, and then retreat to a more secluded part of the enclosure. Many times Jim felt that the size and privacy of the enclosure saved the pack from further injuring or ostracizing Lakota.

Opposite left: Lakota; this page all: Lakota being dominated by Amani.

Beta

Matsi has a special place in the pack as the one who can shift attention away from subordinate wolves that are being dominated or threatened by another member of the pack. Second in size only to Kamots, Matsi has the temperament of a peace-maker. When several wolves at a time chase down Lakota, the omega, he will often follow closely and body-check an aggressive animal out of the pursuit. When new puppies were introduced to the pack, Matsi took on the role of care giver to them.

Opposite page and bottom left: Matsi; top center, left and bottom right: Matsi and Lakota.

Mid-rank & juveniles

Although one wolf may dominate another, in general, dominance may shift during feeding. Motomo is often dominated by his brother Amani except during feeding, when Motomo almost always eats before Amani.

Top right: Amani and Lakota; all other pictures this page: Amani and Motomo. Opposite top: Mid-ranked Amani, dominating Lakota; opposite bottom: Mid-ranked Amani dominating mid-ranked Motomo.

Dominance

Dominance battles and aggression within the Sawtooth Pack escalate at certain times of the year. During the late winter breeding period tensions run highest. Vocalizations increase as do aggressive interactions and conflicts. Dominant wolves more regularly draw blood, and the pack will attack a low-ranking wolf, tearing loose pieces of fur and piercing the skin. Usually these attacks center on the rear and flanks of a subordinate wolf. At other times of the year, dominance battles may be limited to postural threats, stare-downs, and growling.

Opposite: Kamots and Aipuyi; top and center left: Kamots; below left and right: Kamots and Matsi.

Submission

Pups often test the bounds and limitations of adult wolves. They do this to solicit attention as well as to determine their position within the pack. The adults will tolerate this behavior for just so long before they reprimand the pups into submission. What may look to us like a painful correction is usually only display.

Opposite: Lakota; this page: Kamots and Chemukh.

Affection

Lakota, the omega, and Kamots, the alpha, are brothers at opposite spectrums of the pack hierarchy. While they may posture between dominance and submission, displays of affection take place frequently, reaffirming the bonds between them.

This page: Lakota and Kamots; opposite: Lakota, Matsi and Kamots.

Reassurance Displays

Lakota as the low-ranking omega is often mobbed by other pack members. The pack does this to exhibit dominance and to remind Lakota of his position among them. Mobbing is common but rarely fatal. In the Sawtooth Pack, Matsi, the beta wolf, and Lakota's best ally, is often the one to break up the mobbing and offer a comforting lick to Lakota.

Opposite top: Kamots and Motomo; opposite bottom: Motomo, Wyakin and Lakota; this page: Lakota being mobbed by other pack members.

Sparring

Sparring is often spontaneous. It may happen suddenly and for no apparent reason. Kamots and Matsi are close, often showing affection for each other and lying down to sleep together. Despite this allegiance, Kamots periodically forces Matsi to submit. Matsi's submission serves to reinforce the position of both members in the pack.

This page: Kamots and Matsi.

Allegiances

Although wolves have their own social status within the pack, allegiances form and bonds become especially strong between certain individuals. Matsi is the only wolf that spends much time with Lakota. Kamots and Chemukh are pair bonded. They mated in the late winter of 1996.

Pack Hierarchy

Affection, reassurance displays, allegiance, dominance, submission, and sparring are all part of the pack's social structure.

Opposite: Kamots displaying dominance; top left: Kamots and Lakota exhibiting a reassurance display; top right: Motomo submitting to Kamots; center left: Kamots and Motomo wrestling Lakota; bottom: Wahots as a pup.

Play

Play is a common activity among wolves. The Sawtooth Pack plays often between themselves, with objects, even with the crew.

Play is a vital part of every stage of a wolf's life. Wolf puppies roll over each other, scrambling over their siblings in pre-conscious play and dominance assertion. Even before their eyes open, some pups will tend to be the ones that reach their mother's teats first, to end up on top of the heap, and to explore the boundaries of the den. Individual personalities have a determining role in life and are expressed and exhibited during play. As pups grow, play becomes the primary outlet for aggression, hierarchical testing, and exercise. Puppy muscles are toughened and reflexes are toned during bouts of wrestling, chase, digging, and chewing. 🐾 Eventually, play becomes more hunting oriented and skills are developed early that will be used for the entirety of their lives in solo and group hunting activities. As one pup lies asleep in sun-warmed soil, another may be stealthily stalking him, learning the advantages of being downwind. As all siblings play as a group, objects such as sticks, branches, and an adult's tail become prey that the young group learns to subdue. Cooperative hunting skills and social bonding are practiced from the first. 🐾 As the pups put on weight and grow rapidly during the spring and summer, play becomes an arena for testing their strengths and weaknesses. The wolves need to know these things about each other, because when the pack attacks a large animal, or is confronted with danger, it must act quickly and in perfect sychronization. When a 100-pound wolf clamps its jaws on the nose of a half-ton moose and then grimly hangs on while

the moose tries to shake it off, or smash it with it's antlers or hooves, there cannot be any question or hesitancy about what the other wolves on the hunt must do, which is to attack the moose from behind and injure it badly enough to bring it down. 🐾 Play gives wolves the opportunity to mitigate and dissolve the barriers and tensions generated by the strict hierarchical organization of the pack. It is in play that wolves most often temporarily reverse roles. Omega wolves can solicit play from dominant animals. The affinities that develop between wolves of disparate status sometimes develop into friendships and special relationships. Lakota often postures to Matsi and the two will wrestle, then tear across the meadow in full chase. Their friendship is such that Lakota is able to approach Matsi in play even during feedings, which is extraordinary for an omega and a beta. Their relationship, however, does not prevent Matsi from asserting dominant status; one of Lakota's deepest flank wounds was inflicted by Matsi. No one knows if Lakota transgressed some boundary in behavior, or was bitten at Matsi's whim, but their friendship still remains. 🐾 Wolves will play by themselves. Kamots is often seen chasing his own tail, spinning in a circle faster and faster until suddenly, he lines out to chase the nearest pack member into the trees. When feeding, Lakota will sometimes stop eating to play with a piece of hide, tossing it into the air, catching it with his mouth, and then shaking it vigorously. Sometimes he takes a leg bone and twirls it in the air like a baton.

Even though the Sawtooth Pack is familiar with people and tolerates the presence of the film crew, they have maintained their wild identity and characteristics. They are not tame animals.

Running

Games of play and chase develop skills for the hunt. Wolves can run up to 30 or 40 mph when chasing down prey, and cover an average of 30 to 40 miles a day in search of new prey. They have been reported to travel up to 125 miles in a day.

Lakota is often the instigator of play, especially the game of chase. Jostling, tail biting and nipping are all part of the game. For the Sawtooth Pack, play is usually and everyday occurrence. The only time they did not play was during the six weeks after Motaki, then the pack omega, was killed by a mountain lion.

Playing with Objects

Wolves are curious by nature and love to play. They will play with any-thing that excites their interest from hats, to sticks, to old bones. They play tug-o-war, and keep away. In this way they are very similar to their cousin the domestic dog.

Top right: Wyakin, Wahots and Chemukh as young juveniles jockeyed for a piece of the stick; center: Kamots plays tug-o-war with Wyakin and Wahots. This helps to strengthen the puppies jaw muscles for tearing food; bottom left and right: siblings, Wyakin and Wahots are friends one minute, combatants the next; opposite: after playing with a bone Amani gnaws on it to get at the marrow.

Bonding

Although they accept the presence of the film crew the wolves of the Sawtooth Pack are not pets. The pack has maintained its wild behavior because of the careful relationship that Jim has built with them. They are socialized to people, but behave as a pack of wolves with their own rules.

Nurturing

Spring is a time of birth and rebirth in the Sawtooth Mountains. Wolf pups are usually born in late spring when the snows begin to melt and the wildflowers begin to bloom.

Typically, only one pair within a pack breeds and it is often, but not always, the alpha pair. The alpha male may choose to allow another male to mate with the alpha female and this does not jeopardize his role as leader within the pack. This helps ensure more genetic variability within the pack over time, especially if an unrelated male mates with the female. ❧ Both within a pack and within a region, wolves are able to delicately balance population size by interaction in an ecosystem. Wolves breed once a year, in late winter or early spring. Occasionally a pack will forego breeding, often because of pressures from neighboring territories. If prey is scarce and other packs in the area are large, not having a litter of pups increases the pack's chances of survival. If prey is abundant and the territory is extensive, a large litter can be expected. If heavy trapping or hunting pressures take large numbers of a pack, but there are still enough animals to maintain the pack structure, it is likely that the next litter will be large. ❧ After a litter is born, the mother remains in the den with the pups as she nurses them. Soon, she begins to leave the den and the rest of the pack become partners in caring for the new litter. The father and the rest of the adults will come to the den and regurgitate partially digested food for the pups. This is a time when the family aspects of wolf life are most apparent, when the entire pack puts energy, time and food resources into raising and caring for the litter of pups. ❧ When an adult wolf returns to the den or rendezvous site, the pups run to the adult with tails wagging, they face-lick, nip and tug at the adult's face. This greeting results in regurgitation by the adult for the pups. This same greeting between two adult wolves with a

subordinate face-licking and nipping the dominant wolf, no longer results in regurgitation. ❖ When three new pups, Chemukh, Wahots, and Wyakin, were introduced to the Sawtooth pack, Matsi claimed the care of the pups as solely his role. Lakota had been the primary caretaker of Matsi, Amani, and Motomo, and three years later, Matsi guarded the new pups from the rest of the pack. A role as a sole caretaker is unusual for a wolf, but may have benefits within a pack. If Matsi were to become the alpha, the early establishment of an allegiance with the new pups could be useful. As roles change in a wolf pack, individuals remember how they were treated by an ascending animal, and treatment is reciprocated when a status shift occurs. ❖ During early summer, rendezvous sites are established, where the pups are left while the pack hunts or maintains its territory. One or more adults usually remain with the pups to care for the litter until the rest of the pack returns. The role of these adults is to protect and teach the pups as they explore the environment of the site. They are vigilant, even while napping, in watching the movement of the young animals. The pups are typically relentless in their hunting, stalking, charging, and attacking the patient adults. Discipline is firm but gentle and the guardian adult is responsible for containing the energies of the juvenile wolves. The adults are usually very tolerant of the chewing, tugging, and biting that the pups inflict on them but when necessary, will subdue a particularly violent attack. For the pups this roughhousing is a way for them to learn about and test the limits of dominance and skill as well as submission and social boundaries.

Like the second litter, the third litter of pups was adopted from a captive pack in Montana. Before joining the Sawtooth Pack, they were socialized to Jim and the crew. Christina, Jim's daughter, spent a great deal of time with the pups during this crucial bonding period.

Adults respect the rights of the pup's territories, considering found objects and food to be the property of an individual, regardless of status or age. Once a wolf has food within its personal space of approximately a foot, possession is not disputed, not even by a dominant wolf over an omega, unless the food was stolen in the first place. With pups, this respect carries even farther and pups are allowed to take food and keep it away from all adults. ❧ Matsi guarded food for the pups and hoarded it himself to later share with the pups. Once, he became quite aggressive and protective over a pile of meat that he acquired. Any pup could come into this cache and take what it wanted, but no adult could come within several meters of Matsi without a challenge. Even after the pups were fed to near bloating, Matsi would continue to solicit their attention and regurgitate food for them. He continued to guard their food and the remainder that he had collected for days after a feeding. As the pups aged, several other adults occasionally regurgitated food for them but their roles were primarily as teachers and companions. ❧ Kamots demonstrated his dominance over the juveniles to assert his position in the pack and left it well understood that he was the leader. He was also playful with the juveniles and gentle in his discipline. When some infraction occurred, Kamots would generally hold the pup's muzzle in his mouth and pin the pup to the ground for a moment. A growl became enough to suffice and eventually staring down a pup indicated disapproval. ❧ In September of their first year, the pups were nearly full size, although they had not yet reached their adult weights. Every animal in the Sawtooths, including the wolves, was beginning to prepare for a long, cold winter. Their activity level increased dramatically. Mice, voles, squirrels, and chipmunks looking for a haven often found a temporary paradise in Jim's yurt or tent. Those that remained in the tall seed grasses outside were often the unwitting focus of intense chases by the pack. It was a rare event for the wolves to get more than a good smell and fleeting sight of a chipmunk escaping up a tree, but when a rodent carelessly got too far from an escape route, the wolves went into full hunting mode. ❦ One afternoon, the pack was seen wagging their tails excitedly in a circle around a willow bush. Kamots crouched and dove in to retrieve a chipmunk. He carried the rodent over to where the pups were grouped and let the chipmunk go on the ground. As it scurried for cover, the pups saw the movement and lunged. Too late. Kamots again went in after the chipmunk and again returned with it to the pups. This time, he held it near them. When their interest was focused, he let it go again. It easily ran out of reach of the pups. Kamots retrieved it and shook it again, holding it close to the pups. After a few minutes of shaking and displaying, he released it and stood back as it ran in front of the pups. With a paw, he stopped it and again shook it in his mouth, this time to walk slowly away from the pups. As Kamots ate the chipmunk, the pups took turns trying to steal it. ❦ This type of learning situation typifies how a wolf pack integrates social bonding, teaching, and play. As the adults with a pack guide younger animals with hunting skills, bonds are formed, roles defined, and the continuous process of behavioral displays reinforce social standing and communal hierarchy.

Nurturing is an important part of the bonding process between the pups and the crew, just as it is between the pack members.

While perhaps only half the pups born in a litter will survive to their first year of age, succumbing to predation, disease, and natural attrition, the remainder will have learned a substantial number of survival skills and will most likely live to adulthood. An old wolf in the wild might live to be ten, eleven, or maybe twelve years old. It will face constant challenges for acquiring food, companionship, and territory. Wolves prey primarily upon deer, elk, moose, and caribou. Few wolves are left unmarked by their prey. The kick of a deer's hoof can easily cave in a wolf's skull or break a large bone. The kick from an elk is more powerful and that of a moose more powerful still. Packs hunt cooperatively for many reasons. One benefit of having several wolves pursuing one prey animal is that a wolf that is part of a pack may more easily evade a kick or charge from a large ungulate.

Living among the wolves in a yurt that elevated seven feet above the ground allowed Jim and Jamie to watch, film and record the wolves without interfering with their natural behavior. The nurturing nature of wolves is well-known but rarely captured on film in the way that Jim and Jamie were able to document it.

Feeding

Wolves are usually extremely protective of the prey that they have killed. Very little will induce them to surrender their food, except for bears and humans. Because of the unique bond that Jim and Jamie developed with the Sawtooth Pack, they were able to observe and film them in their relaxed state.

Competition and predation are intertwined in the animal kingdom. The coyote is a competitor of the wolf, preying on many of the same animals that wolves do. Wolves have been known to kill coyotes to protect their territory. Beaver, deer, rabbit and elk are just some of the animals that wolves prey upon.

Wolves eat as much as 25-30 pounds in a feeding, then may go for days without eating again. Nothing is wasted. Wolves eat almost every part of their prey including the meat, internal organs, hooves and bones, which are crushed to get at the marrow.

The wolf is the only major predator in North America that hunts in a pack. Coyotes will occasionally hunt in packs but this is less common. By hunting in packs, wolves maximize their chances of a successful kill and minimize their chances of injury, especially when they are hunting big game animals like elk and moose.

Nurturing

At about the age of 12 weeks the pups were old enough to join the rest of the pack. The adult wolves immediately accepted them as their own. It was an exciting time. During the first year, the pups were always under the watchful eye of the adults with at least one wolf serving as the primary care giver.

Communication

The howl of the wolf is legendary. For centuries, if not millenniums, it has stirred us to passion. So much so, that we often forget that wolves communicate in many different ways. Eye contact, scent marking, vocalizing, facial movement and body language are just some of the ways that wolves express themselves.

Wolves communicate on many levels, both overt and subtle. Many forms of communication involve smells and sounds beyond the range of human hearing. Many postural and facial expressions are so subtle that only people who have lived with wolves are aware of them, and even then, many slight gestures go undetected. ❦ Physical postures are continuous reminders of social status among pack members and often indicate mood as well. As most dog owners can attest, a dog's body postures are the clearest form of communication between dog and owner. A wagging tail, ears alert and up, a straight back, and eyes bright, are undeniably characteristic of a happy dog. Two dogs interacting for the first time are often excellent examples of postural communication. Dominance, submission, exploration, and greeting are gestures we are familiar with in observing our pets. These postures are "wolf-like" and indicate familial links between wolves and dogs. Wolves tend to have a more elaborate repertoire of postural gestures, necessary to a member of a species interacting with other members on a constant basis. ❦ Some examples of dominance postures, from most threatening to least are: standing over a subordinate with hackles raised, ears flattened, and teeth bared; hackles slightly raised, low growling, while staring a subordinate down; low growling; staring determinedly at a subordinate; ignoring a subordinate, and turning away. ❦ Some examples of submission from most submissive to least are: rolling over onto the back with tail curved up over the belly, urinating, whining; rolling onto one side, paw raised, lips curled back, whining; sitting with one paw raised, ears flat, licking at the face of a dominant animal;

haunches tucked under with tail curled up underneath, ears flattened, eyes downcast; tail low and wagging; and eyes averted. ❧ Some examples of friendliness are: tails wagging, heads low; smelling and snuffling at each other's mouths and faces; jumping up with forelegs circling each other's neck regions as a solicitation for play; play. ❧ Vocal communication in wolves is fascinating and far from understood by humans. Vocalization categories include squeaking, a high-pitched, short sound; and whining, which is also high-pitched but lasts longer and can include undulations. These sounds are used by submissive wolves and in greetings or when a wolf wants something it cannot reach or obtain. Barking is used as a threat, challenge, or warning. Bark-howls are often a warning to the pack of perceived threat. Growling is used as a warning primarily among pack members by more dominant animals but is also used as a warning of intruders. ❧ Howling is the most famous wolf call and is quite complex. Howling can range from around 100 to 800 cycles and has up to or exceeding twelve harmonically related overtones. Contrary to popular belief, howling does not occur more frequently during full moons. While howling does aid wolves in assembling the pack for hunting, outside challenges, and socialization, there are infinite reasons why wolves howl. There seem to be welcoming howls when a pack member or familiar person (in a captive situation) returns, howling before feeding or hunting, and howling to advertise territory. Howling seems to inhibit intruders when that territory is also boundary marked by scent. Researchers now believe that howling and other vocalizations carry individual pieces of information that

characterize an animal by a spectrograph or voice print. This voice print may eventually become useful as a viable method of identification. It is also likely that a voice print could be read to understand something about social status, since more dominant animals tend to have greater overtones and richer vocal repertoires. Pups also have distinctive vocalizations and dramatize that difference when threatened by an adult. 🐾 Smell is another avenue of communication that is little understood. Wolves have a much keener sense of smell and larger olfactory lobes in the brain for processing information. Scent glands are located along their sides and on their faces. They have anal glands and a post-caudal scent gland at the dark spot on their tails and between their toes. Wolves will often mark an object by rubbing against it and concurrently carry away the scent of that object. Wolves are attracted to strong odors and will roll and scent mark in areas where urine or scat have been left. Dominant animals scent mark most frequently and the alpha animals most of all. 🐾 Scent marking the boundaries of a pack's territory seems to be used by a pack to define its own boundaries as much as to advertise to outsiders where the lines are drawn. Scent marking is usually accomplished by urination, defecation, or by rubbing the scent glands. Scratching the ground, especially after urination, further emphasizes the marking.

Wolves communicate with each other to express many different emotions including dominance, reassurance, pleasure and anger. All forms of communication serve to reinforce the powerful bond of the pack.

Wolves howl for many reasons. The first howl is often initiated by the alpha leader. Sometimes it is spontaneous. Sometimes it is with a purpose: to call the pack together, reaffirm territory, or respond to far off sounds. Each wolf has it's own distinctive howl. When wolves howl together it sounds like a chorus. Howling usually occurs late at night and in the very early morning hours when it is still dark. However, wolves will howl during the day.

The communication system of wolves is highly developed and complex. While we may never completely understand the wolf, this project has brought us one step closer to knowing them better.

Opposite: Lakota and Matsi have a special bond; top left: Wyakin in an act of submission to Matsi, begs for permission to feed; top right: Chemukh bares her teeth to appear aggressive, but her flattened ears indicate fear and subordination; center left: a tucked tail indicates submission; bottom: Matsi communicates dominance over a terrified subordinate, Chemukh.

A New Home

Nearly a century after wolves were almost completely eradicated from the western United States they are finally being returned to central Idaho and Yellowstone National Park. Their future is still uncertain and will depend greatly on whether or not many long-held beliefs and assumptions about them can change.

During the summer of 1996, the Sawtooth Pack was moved to the Nez Perce reservation in west-central Idaho. What will remain a permanent home for the Sawtooth Pack is the culmination of efforts on behalf of the Wolf Education and Research Center, an organization that Jim Dutcher founded because of his concern for the wolf. The Nez Perce tribe has provided 300 acres of their tribal lands for the Wolf Education and Research Center to build a new visitor's center and enclosure for the pack. The center will provide the public with a rare opportunity to learn more about the wolf in its native habitat, and to experience the rich cultural and natural history of the Nez Perce Indians. ❧ The Wolf Education and Research Center has been active in wolf education programs primarily throughout rural Idaho, reaching school children in an attempt to dispel myths and untruths about this large predator. Research has been ongoing with the Sawtooth Pack on behavioral and vocalization aspects of wolves. The Wolf Education and Research Center has also been instrumental in helping to re-establish wolf populations in the northern Rockies, specifically in Idaho. ❧ Currently, the Nez Perce have jurisdiction over wolf management in Idaho. In 1995 and 1996, two groups of wolves were trapped in Canada and relocated to wilderness areas in central Idaho and in Yellowstone National Park. The wolves were flown down to the United States.

The animals slated for Yellowstone were released into large pens within the park, where they were fed and habituated to their new home for several months before their release into the wild. In Idaho, the wolves were released directly from their transport crates into federal wilderness. In 1995, 29 wolves were returned to Yellowstone and Idaho. In 1996, 37 more were brought to these locations. ❖ The return of the wolf is the return of a vital component of ecosystems. In places like Yellowstone National Park, herds of bison, elk, and deer have outgrown the capacity of the habitat. As a result, many of them starve to death every winter. It is hoped and believed that returning the wolf to Yellowstone National Park will return some balance to the ecosystem by keeping ungulate populations in check. Only time will tell. ❖ When he began this project in 1990, Jim always knew that the Sawtooth Pack could never be set free. He also knew that one day his special permit with the U.S. Forest Service for the use of their land would expire and that the Sawtooth Pack would have to be moved to their permanent home. Carla Higheagle (opposite page) came to visit the pack before they were moved to the Nez Perce Reservation. She spoke during her visit of the interconnectedness that she felt between the wolves and her people, the persecution they both experienced, and the bright hope they now share for the future.

When the time came to say good-bye, and for the wolves to leave for their new home, Jim knew that he would miss them and that perhaps they would miss him, too. They had developed a special bond, a bond between between man and wolf, that few people will ever know.

Appendix

Wolves Are Not Pets

Many people wonder: would it be possible to be friends with a wolf? There are many ways to be friends with a wolf, or with wolves in general, but these do not include having a wolf as a pet. ❧ If you raise a wolf from an early age—and owing to the nature of a wolf this is the only way of having one bond with you—you will probably fall in love with it. At the least, there will be closeness and appreciation. A relationship with a wild animal is deeply exciting. It is also an almost certain appointment with terrible heartbreak. ❧ A wolf is a wild animal. If you make a wolf into a pet, you make it dependent on you. This is against the basic nature of a wild animal, and because this wolf has been socialized to humans, it can never live a natural life again. If released into the wild, it will not know how to get along or how to behave. Sooner or later it will die—probably from a bullet fired by a frightened or angry human. ❧ When you make a wolf a pet, you are stuck with each other. Wolves, when they are puppies, are totally endearing. Wolves when they reach sexual maturity at about three years of age can become unpredictable and very difficult to live with, even though they are socialized to a human. ❧ YOU CANNOT TRAIN A WOLF! Repeat: YOU CANNOT TRAIN A WOLF! Dogs can be trained. This is because they have been domesticated, selectively bred for thousands of years for traits of obedience and cooperation. Most dogs enjoy training. ❧ A wolf simply goes its own way. A wolf will not respond to a name, or to commands. If something inside a couch merits investigation, a wolf will take the couch apart with its teeth. In a few months, what starts as a charming, endearing puppy becomes a 100-pound, incredibly strong dynamo that must live in a special place where damage can be kept to a minimum. ❧ Wolves have complex behaviors. It is difficult in the extreme for a human being to understand a wolf, and to not at some point have a conflict as a result of a misunderstanding. In the case of strangers and people unfamiliar with animals, the possibility of an incident is far greater. So as the pet wolf becomes older, contact with other humans has to be carefully monitored. This increases the isolation and alienation for an animal that is fundamentally social. ❧ A wolf needs space. Creating a strong, safe enclosure of any reasonable size is very expensive; a fence for 20 acres will cost around $200,000. Once constructed, the fence has to be carefully monitored and maintained. ❧ In most cases, would-be wolf pet owners bow to the inevitable. They realize that wild creatures do not belong in human society, and often, they try to give their wolf to an animal sanctuary. There are perhaps hundreds of small, struggling wolf rescue centers, devoted to caring for unwanted wolves and other animals. But they have limited capacity. Caring for a wolf is expensive. So many times, an unwanted wolf—that wonderful cute puppy, that life-long soulmate, now a great, powerful creature who still licks you and trusts you with all of his or her heart—is simply put down. Killed. ❧ That's the usual script, and yet astonishingly, there is apparently a flourishing trade in wolves as pets. Wolves are an endangered species, and it is against the law to interfere in any way with a wolf, much less trade in wolves. However, if a wolf mates with a dog, the result is a hybrid no longer classed as a wolf. A hybrid that is just 3 percent dog and 97 percent wolf is excluded from protection as an endangered species. ❧ The classified ads in various wolf and exotic pet magazines are full of offers for '97%' hybrids. What the animals are in terms of genetic make-up is anyone's guess. (Many are more dog than wolf.) But any hybrid that is truly '97%' is a wolf, no matter what the rules say. ❧ Owning a wolf or a hybrid has apparently become something of a craze. Yet a wolf is a social animal. It needs other wolves to live with. It is cruel and

upsetting to deny a wolf the companionship of its own kind. ❧ A wolf can be a wolf, and still be friends with humans. However, it is not a casual business. Captive wolves, accustomed to humans, can be unpredictable, and great care and sensitivity must be exercised when interacting with them. It is really easy for something to go wrong. Wolves are very complex, and a wolf cut off from other wolves, living in circumstances that are peculiar and unusual for a wolf, may be truly unpredictable. Remember, although wolves are very sensitive to humans, they do not learn our language or, like the dog, seek to cooperate. Rather, it is up to us to learn the language of wolves. And even the best, most experienced wolf biologists and handlers sometimes have problems. Someone who simply has a hazy, romantic idea that they will 'get along' with a wolf is almost certain to at some point have an unsettling or even dangerous experience. Life with dear wolf is thereafter never the same. ❧ It is true that there are some wolves who seem to genuinely enjoy human company. They are, within limits, 'safe' to take to community centers, schools, and other social gatherings. But wolves are individuals. There is no way to predict, out of a litter of pups, if a particular wolf will grow up to be well-adjusted to humans—or to be fearful or aggressive and prone to bite. Yet once a wolf has been socialized, there is no turning back. ❧ The only reasonable way to keep wolves in captivity is as a pack. The wolves then have at least some of the elements of a normal wolf life—the entity of the pack, which is the wolf, and some room in which to move around. So far as human interaction is concerned, it has to be restricted to a few people who respect wolf language. ❧ The friendliest, most beneficial thing to do for wolves is to work towards a wild environment where they can live as nature intended. Captive packs and wolves in zoos can help teach people that wolves are not the monsters of myth and have every right to live. But the only thing that will save the wolf is space—wilderness. ❧ Our wilderness is diminishing and shrinking. We all know this. The only question is: how much wilderness do we preserve? And, will it be enough? Support the wolf by all means. But even more, support the efforts to preserve the wild habitats in which wolves, and many other animals,

can live. There are many organizations working to save the wolf (see
Resources). They can all use your participation and help. 🐾 Support also, at
every opportunity, a change in our attitudes. There's no need to fear wolves,
or wipe them out. In Minnesota, over 2,000 wild wolves live without undue
conflict with humans. If a particular wolf takes it in mind to consistently prey
on livestock, any damage done is compensated, and that wolf is culled. That
may sound drastic, but so far, the cost of damages has been modest, and only

a few wolves have had to be eliminated. ❧ Each area of the country is unique and what works in Minnesota might not do so well in say, Wyoming, and might not work at all in New Jersey. But wolves are adaptable, and smart. They can learn, too. If we but try, and make well-informed efforts to coexist with wolves, then in a surprising number of places, we may well see what is natural and right—the return of the wolf.

—RICHARD BALLANTINE

Resources

Wolf Organizations

Wolf organizations vary considerably in scope and character. Some are large and are involved in ongoing wolf research and wolf reintroduction programs, others are small and devoted largely to the care of animals. Many offer a variety of interaction points, including newsletters, adopt-a-wolf programs, and sometimes, it is possible to do volunteer work—for example, fence-building at a sanctuary. Write for details, and please be sure to include a self-addressed, stamped, large envelope.

Alaska Wildlife Alliance
PO Box 202022
Anchorage, AK 99520

Ancient Forest Rescue
483 Marine Street
Boulder, CO 80302

Canadian Wolf Defenders
PO Box 3480, Station D
Edmonton 50, Alberta T5L 4J3
Canada

Clem & Jethro Lectures
PO Box 5817
Santa Fe, NM 87502

Defenders of Wildlife Wolf Fund
1244 19th Street NW
Washington, DC. 20036

Endangered Species Coalition
666 Pennsylvania Avenue SE
Washington, DC 20003

Friends of the Wolf
PO Box 21032
Glebe Postal Outlet
Ottawa, Ontario K1S 5N1
Canada

HOWL
4600 Emerson Avenue South
Minneapolis, MN 55409

International Wolf Center
1369 Highway 169
Ely, MN 55731-8129

Maine Wolf Coalition
RFD #6
Box 533
Augusta, ME 04330

Mexican Wolf Coalition
207 San Pedro, NE
Albuquerque, NM 87108

Mission: Wolf
PO Box 211
Silver Cliff, CO 81249

Montana Wilderness Association
PO Box 635
Helena, MT 59624

National Wildlife Federation
240 North Higgins #2
Boulder, CO 80303

Predator Project
PO Box 6733
Bozeman, MT 59771

Preserve Arizona's Wolves
1413 East Dobbins Road
Phoenix, AZ 85040

RESTORE: The North Woods
PO Box 440
Concord, MA 01742

Red Wolf Recovery Fund
National Fish and Wildlife Foundation
18th and C Streets NW
Washington, DC. 20240

Red Wolf Fund Tacoma Zoological Society
5400 North Pearl Street
Tacoma, Washington 98407

Sinapu
PO Box 3243
Boulder, CO 80307

Timber Wolf Alliance
Northland College
Ashland, WI 54806-3999

Timber Wolf Preservation Society
6669 South 76th Street
Greenside, WI 53129

WOLF!
PO Box 112
Clifton Heights, PA 19018

Wolf Action Group
2118 Central SE
Suite 46
Albuquerque, NM 87108

Wolf Action Network
PO Box 6733
Bozeman, MT 59771

Wolf Awareness
RR #3
Ontario N0M 1A0
Canada

Wolf Ecology Project
School of Forestry
University of Montana
Missoula, MT 59812

Wolf Education Fund
Zion Natural History Association
Springdale, Utah 84767

Sawtooth Wolf Pack Adoption Kit

Wolf Education and Research Center
PO Box 3832
Ketchum, ID 83340
Home of the Sawtooth Wolf Pack

The Wolf Fund
PO Box 471
Moose, WY 83012

Wolf Haven International
3111 Offut Lake Road
Tenino, WA 98589

Wolf Hollow
Route 133
Ipswich, MA 09138

Wolf Park
Battle Ground, IN 47920

Wolf Recovery Foundation
PO Box 793
Boise, ID 83701

Wolf Sanctuary
PO Box 760
Eureka, MO 63025

Wolf Song of Alaska
6430 Ridge Tree Circle
Anchorage, AK 99516

MAGAZINES

Soul of the Wolf
World Wildlife Sanctuary
PO Box 1026
Agoura, CA 91301
Quarterly

Wolf!
PO Box 29
Lafayette, IN 47902-0029
(E-mail) Wolfpark@aol.com
Quarterly

Photo Credits

Key:

L-left   T-top
R-right   B-bottom
TL-top left   BL-bottom left
TR-top right   BR-bottom right
CL-center left   CR-center right
FP-full page

Introduction

p.7 BR: Jamie Dutcher
p.10 TL: Franz Camenzind
p.10 CL: Jamie Dutcher
p.11 TR: Johann Guschelbauer
p.12 TL: Jake Provonsha
p.14 T: Janet Kellam
p.15 TR: Jamie Dutcher
p.15 3TR: Shane Stent
p.17 BR: Shane Stent
p.18 FP: Jamie Dutcher

The Project

p.24 1L: Franz Camenzind
p.24 2L: Franz Camenzind
p.24 3L: Janet Kellam
p.24 4L: Janet Kellam
p.30 TL: Janet Kellam

p.32 BL: Bob Poole
p.34 1L: Shane Stent
p.34 2L: Shane Stent
p.35 FP: Shane Stent
p.36 1L: Franz Camenzind
p.36 2L: Janet Kellam
p.36 3L: Bob Poole

Family Life

p.50 All: Jamie Dutcher
p.59 TC and LR: Jamie Dutcher
p.78 All: Dutcher
p.85 TL: Jamie Dutcher

Play

p.91 2T: Jamie Dutcher
p.91 3, 4T: Shane Stent
p.91 5T: Jamie Dutcher
p.100 TL: Janet Kellam
p.101 L: Jamie Dutcher
p.101 CR: Jamie Dutcher
p.101 BL: Jamie Dutcher
p.101 BR: Jamie Dutcher
p.106 CL: Jamie Dutcher
p.108 B: Jamie Dutcher

Nurturing

p.113 1T: Shane Stent
p.115 1T: Bob Poole
p.115 3T: Bob Poole
p.116 T: Jamie Dutcher
p.120 2TR: Johann Guschelbauer

Communication

p.144 3T: Jamie Dutcher
p.150 FP: Johann Guschelbauer
p.155 T: Jamie Dutcher
p.157 B: Jamie Dutcher

New Home

p.162 3T: Jamie Dutcher
p.162 4T: Shane Stent
p.163 2T: Shane Stent
p.171 B: Jamie Dutcher

Appendix

p.190 T: Jamie Dutcher

Acknowledgment

There were many people who helped me with the Sawtooth Pack project. None of this would have been possible without them. More than anyone, I would like to thank my wife and companion, Jamie. She was always with me to support me and to offer a helping hand with everything from the politics to planning to logistics, to puppy raising to photography to sound recording. I could not have done this without her. ❧ I would also like to thank the people who worked with me on "Wolf: Return of a Legend," which basically got this project off the ground. It would be impossible to thank everyone here but I would especially like to thank Janet Kellam, my associate producer, who was involved from the beginning when this was just an idea; Dennis Kane and David O'Dell who had faith in us to put it on national network television; and Bob Poole, Franz Camenzind, Jake and Patty Provonsha, Jan Roddy, Lisa Lundal Diekmann, Burke Smith, Sara Bingaman, Garrick and Christina Dutcher, who also played an instrumental role in the making of the movie. ❧ This project wouldn't have been possible without the dedication of many people who helped care for the wolves including Susie Weere, Lori Schmidt, Val Asher, Mark and Missy Fitzgerald, Gary Gadwa, Brent Snider, Jerry's Country Store; and especially Megan Parker who joined the wolves when they moved to the Nez Perce Reservation. ❧ Many other people provided assistance to us including Piero and Mable Piva, and the Furey and Racine families; who allowed us and the wolves to be their neighbors; Randy Acker and Grant Mauer provided essential veterinary care; Jan Wygle and Kathy Stetina; Karin and Neil, two very special friends, provided the pups to us for adoption; Diana and Mallory Walker and Teresa Heinz for their advice and counsel. ❧ Lastly I would like to thank the people who helped me with the book: Megan Parker, for all the research that she did as well as for outlining the eventual text and for the wonderful way she has about telling our story; Lisa Lundal Diekmann for coming to our rescue at the last minute and for writing some of the captions; my father, Corneil Dutcher for his help with the Introduction, Vanessa Schulz and Page and Maureen Jenner. Finally my long time good friend, Johann Guschelbauer and Shane Stent for helping with the photography.

—JIM DUTCHER